Sarah Schoonmaker Baker

Fireside Sketches from Swedish Life

Sarah Schoonmaker Baker

Fireside Sketches from Swedish Life

ISBN/EAN: 9783337253158

Printed in Europe, USA, Canada, Australia, Japan

Cover: Foto ©berggeist007 / pixelio.de

More available books at **www.hansebooks.com**

FIRESIDE SKETCHES FROM SWEDISH LIFE.

CHRISTMAS FARE.

Page 33.

Fireside Sketches

FROM

Swedish Life

T. Nelson & Sons

FIRESIDE SKETCHES

FROM

SWEDISH LIFE

BY

Mrs. WOODS BAKER

*Author of "The Swedish Twins," "Pictures of Swedish Life,"
"The Babes in the Basket," etc.*

T. NELSON AND SONS

London, Edinburgh, and New York

1896

CONTENTS.

FIRESIDE SKETCHES FROM SWEDISH LIFE.

FARFAR PETER.

I.—THE OLD " EXCEPTION."

FARFAR PETER (father's father Peter) lived in a little red cottage on one of the great plains of Sweden. He looked much like a gnarled old oak, still sturdy, but damaged and battered by the storms of many long years. There was, indeed, a bit of oak which was in a way a part of himself ; for it took the place of one leg, from the knee down, and was a solid and tolerably serviceable, if not graceful, member. A falling tree had long ago crushed what Peter now called " my buried leg," and of which he made mention as respectfully as if a whole volley of shot had been fired over it as a funeral salute.

"The year I buried my leg" was a date frequently alluded to by Farfar Peter, and naturally, as it was a changing point in his life. Then, when no longer fit for active labour, he had sold his little home, and the bit of land attached to it, to his vigorous son, who moved into the cottage at once, with all his rosy youngsters. Farfar Peter was henceforward to have a right as long as he lived to bed and board and a room to himself in the cottage, with light and fuel provided. So ran the terms of the contract. Farfar Peter was so made a reserve man, a reservation, an exception, as such a person may be called, the Swedish language having a peculiar expression adapted to the peculiar custom. He had exchanged his home for his stipulated provision for life.

Farfar Peter had a cheerful disposition, and willing, skilful hands. He applied himself now wholly to what had before been his occupation for the long winter evenings. So all sorts of wooden implements and vessels for domestic use were fashioned by his dexterous fingers.

Around him there generally lay a soft carpet of white shavings, in the midst of which he sat more contented than a king upon his throne. There his grandchildren liked to gather about him, not as in the shadow of a great old tree, but rather as they would seek the sunshine on a chilly day.

Farfar Peter had had, on the whole, a happy time since the burial of his leg, until his son Axel suddenly told the old man one day that he was thinking of emigrating to America. A former neighbour had written to Axel that there was a grand chance for providing for a young family in that part of the New World where he had settled. In a certain seaport town where he had taken up his abode there was a great factory for the making of india-rubber over-shoes, where the men were paid by the day according to the amount and skilful execution of their work. With "Axel's handy way of doing things," the friend was sure he would make money, and in time he might get a place for his sons in the same establish-ment. In short, Axel had evidently made up his mind for the move. "The fact is," he said, "I can't help seeing that the boys are growing up fast, and what chance is there for them here?" And Axel gave a circular sweep with his long arms that took in a large part of Farfar's little room, and seemed to comprehend every bit of the wide plain without, if not all Sweden.

"Of course you would go too, father," said Axel.

"We could look out for you there as well as here, and better too, likely, from what I hear. I shall sell out, and I think I know a man who will buy; so my pocket will not be empty for the start."

Farfar Peter stood up, and leaned with one hand on the rough table before him. "Leave old Sweden! Leave this cottage!" he exclaimed vehemently. "Here I was born, and here I hope to die. No, no! Go where you please, Axel. Here I belong, and here I mean to stay. I can be 'an exception' for the buyer as I have been for you. Don't think of me!"

"But we do think of you, father, and we want you to go too. Stina and I don't like the move, but its for the children. One must think of the children's future, and there is a better chance for them all over there. It's for the children, you see," urged the son.

"I see! I see well enough," said the old man; "but I expect to close my eyes here where I was born. I am not like women, who must have folk round them all the time, as if they were afraid of their own shadows when they are alone. I can get on by myself, never you fear. I've got my home here made sure, so don't let me hear another word about my going over the water. I'm company for myself, specially when I whistle, as I do mostly when I'm alone."

Farfar Peter had accepted the probable change then and there, and would positively hear no more talk upon the subject. The children one and all set up a wail of remonstrance when they heard that Farfar was to be left behind. He looked at them with grim satisfaction in their rebellion, but was unmoved in his purpose. By degrees they grew accustomed to hear Farfar talk of the nice time he was to have when they were all off over the water.

Axel was prompt in action. The purchaser, a young married man, was glad to have such a snug home in which to put his precious wife, and did not think he should find " Farfar Peter " the slightest inconvenience, as he had a good name the country round. The old man should be dealt liberally with. There should be no scrimping with " the exception."

Axel and his family were soon far away, and Farfar Peter was making his discoveries at home. The young couple he had set down as a pair of cooing fools who could think of nobody but each other.

The day after their arrival, the husband had boarded up the door that led from Farfar's room to the rest of the cottage. He did not need it, they said, as he had

his own entrance to use, and they could come in that
way, too, when they wanted to see him.   It was not
often that they availed themselves of that privilege.
Farfar's housekeeping did not suit the young wife on
her first visit, and she did not repeat it.   The husband
looked in only to bring the supplies that he was
pledged to give to the old man.   Abundant supplies
they were, and more than up to the contract.   It had
not been stipulated in the agreement that Farfar
Peter's food should be cooked—that had naturally been
taken for granted by Axel.   Farfar proved quite com-
petent to boil a dish of rye porridge or bake a potato,
and this came to be the extent of his cooking.   With
raw ham as a luxury now and then, cheese, hard
brown bread, and herring soon made, with milk, his
general bill of fare.   Perhaps Farfar Peter got
dyspeptic in consequence of this uniform diet.   He
made, in fact, before long some unpleasant discoveries
about himself that quite surprised him.   He was not
such a sunny old fellow as he had supposed.   He had
only reflected the cheerfulness that had flashed and
sparkled from the children that clustered around him.
It was plain he had been kept up by his share of the
family life and love.   He was growing bitter, and had
his own dreary reflections.   "Love is like money," he

said to himself; "it goes down in the family, not back. It goes from the parents to the children, not back to the old folks. Axel must be off to America for the children, to be sure. Old Farfar could get on very well in Sweden alone."

Farfar went down in his own estimation. He knew he was getting to be a grumbling old fellow, and he was ashamed of it. When the long autumn evenings came, he shut his teeth firmly in a sort of defiance of the world in general, and especially of the unconscious and happy young couple on the other side of the door.

When it grew dark at last, at four o'clock in the afternoon, Peter learned to count his day as done then, and to go to bed to keep warm and get rid of his dolorous thoughts. So, as the old man had almost a bear's power of sleeping, he managed to dispose comfortably, if not profitably, of a large proportion of his lonely hours.

In dreams he was far back in the sunny past, and was happy.

II.—NEIGHBOURS.

FARFAR PETER'S was not a literary family. A postal card came now and then from America, curtly stating

that all were well and contented in the new home, and
hoping that Farfar was getting on comfortably.

In point of fact, the emigrants were delighted with
their pleasant quarters and their undreamed-of prosperity. Axel had made a bank-deposit on his arrival
of the remains from the sale of his Swedish property,
and had at once hired a small house in the suburbs of
the town, where he had promptly procured work, as
his friend had anticipated.

There was a little garden about the new home,
with narrow, trodden paths here and there, and an
old pear-tree in the corner that the children regarded
as a kind of sacred treasure. Axel and the boys at
once set to work to widen the paths, and cover them
with loose sand many inches deep in their Swedish
fashion. On Saturday evening the whole family
were sure to be out taking turns with the home-made
rakes till every bit of path or open ground bore the
parallel marks that told that the last week's traces
were fairly blotted out, and all made clear and fair to
begin anew on the Sabbath morning. The first foot-
steps on the path from the doorstep were sure to be
those of the whole family on their way to the Swedish
chapel, where they would have their own service in
their own well-beloved language.

Within the house all was as neat as busy hands could make it. The furniture was new and of the simplest sort; the old stock at home had been sold with the cottage. Many warnings the children received not to be pressing their faces against the clean window-panes, but the temptation was irresistible.

Accustomed so long to look out on the broad, Swedish plain, with hardly a human habitation in sight, the little family had now opposite neighbours to inspect and to regard with insatiable curiosity.

Across the wide street there was a house, the match to their own in size and form and surroundings, but quite different in its whole condition and expression. There Pat and Bridget Malony had their home, with a set of noisy, wide-mouthed, merry children, who seemed to be always in circulation from morning till night.

Bridget had one established principle for keeping her house "tight and trim." Was a cup broken? "Throw it out," was the order. For sweepings, apple-parings, old tin pans, old shoes, and indeed all refuse, the command was the same, and sure to be promptly executed. To Bridget's mind all things seemed to disappear into infinite space as soon as

they were fairly cast out from window or door.   She
had no further responsibility in the matter.

Swinging on the gate and gymnastics on the fence
were the favourite pastimes of the Irish boys.   If a
bar or a " picket " were by chance torn off now and
then, it was a joy to jump on it and subdue it to firewood to be triumphantly carried to " Mammy " for
domestic use.

It can easily be understood that these doings were
not according to the established code in the Swedish
home.   There was a general feeling of contempt for
" those Irish " in Stina's family—a form of thankfulness that they were not as their neighbours.
There grew up by degrees a kind of animosity
between the two households, which showed itself
among the Irish flock by open taunts addressed to
the neat-looking children who went in and out from
the enemy's quarters, as clean and stiff, the tor-
mentors said, as if they were choir boys on duty.
The Swedes did not understand these contemptuous
remarks, but the howls and clenched fists which
accompanied them were most intelligible, and induced
in the persecuted a glow of righteous indignation.

There was, at last, a council of war held behind
Stina's house, after which the Swedes, who had been

well trained in gymnastics at school, rushed in a body on the Irish squadrons, and ignominiously defeated them. After having so "conquered a peace," the victorious party affected not even to know that their opposite neighbours were in existence. All the while, however, the window corners were sedulously used as places of observation, and the Swedes were as familiar with "the habits" of their quondam enemies as was ever an enthusiastic naturalist with the wonderful doings of a colony of ants he had taken under his especial protection.

One morning Stina and Axel were hastily called, and stood with their children boldly at the window to see what could be going on at the Irish home.

A handsome covered carriage, with a coachman and a pair of fine horses, was standing at the gate over the way. Pat had gone early to his work, with his shovel over his shoulder as usual, but Bridget, in her gay Sunday attire, stepped into the luxurious vehicle, her face all abeam with satisfaction.

An hour afterwards there was another summons to the window. The carriage had returned, and Bridget was no longer alone. She got out first, and then tenderly helped down a queer, little old man, to whom she gave a strong hug as he set his feet on the

ground, and then administered to him an emphatic
push inside the gate, to intimate that he was welcome
to the premises.   As he stood at the foot of the steps,
the observers could see that he wore a high, tall-
crowned hat above his thin, grey hair, and a long
coat down to the heels of his slender, unstable figure,
and a pair of trousers every way suited to a man of
more pretentious proportions.

A fat little woman, with a blanket-shawl around
her, and a dress as short as her husband's coat was
long, next planted her feet firmly on American soil.
She wore no bonnet.   As she threw back her head
to be kissed, the broad frill to her close, white cap
haloed round her face, and made it look like the sun
in the almanac.

Such a smack as she got from her daughter!
This proceeding was just over, when there was a
yell of delight as the whole Irish brood came storming along the street, jubilating in a wonderful fashion.
Having taken the opportunity for some predatory
expedition during their mother's absence, they were
now returning in a body, and fell on the old people
with kisses and shouts of " Grandpappy ! " and
"Granny ! " that to all English ears told the whole
story.

The story needed no words to explain itself.
When the clumsy little old chest, which the coachman had condescended to put his feet on as he drove,
was fairly deposited, Bridget pompously paid the fare,
not unwilling that the neighbours should see that
more than one solid bit of silver was dropped into
the coachman's open hand.

The transaction was evidently over. Tears stood
in Stina's eyes as she went back to her work. Axel
kept his mouth shut all day, as if he had the tooth-
ache, and as for the children, they did not hesitate to
say out plainly, "What a pity our Farfar Peter
wouldn't come to America with us!" They felt that
the Irish family had now the advantage over them,
and it was hard to be beat. "Think!" they said,
"those boys did not even know they had a grandfather too, and he almost twice as big as that little
fellow with the high hat!"

Other matters soon swept in to take Stina's
thoughts. Her oldest boy had been already able
to get work in the factory with his father, and
with their ingenious Swedish hands, accustomed to
making all kinds of implements for home use, they
were both soon skilful workmen, and able to earn
high wages. The second boy had not yet found any

permanent employment. He seemed drooping, and
finally announced suddenly in the assembled family
circle that he wanted to go to sea. There was a
look of astonishment and dissatisfaction on the part
of the elders, and of respectful admiration from the
children for the pluck that Hans was displaying.

"We'll talk it over with you this evening," said
Axel, wishing to gain time to think how to meet
this revolutionary outbreak. The family council
took place on the door-steps that evening—a council
of three, the parents and young Axel—the rest of
the children being variously disposed of, according
to their respective ages. The result was a general
surprise: Hans was to have his way, and to sea he
went, when the autumn leaves were falling.

III.—IN THE RED COTTAGE.

CHRISTMAS was coming. Farfar Peter knew it, but
there were no signs of preparation in the cottage.
No rose-coloured pig was hung on the spike in
the trunk of the great ash-tree that overshadowed
the humble home. There had been no sound of
chopping or pounding or beating of eggs on the other

side of the wall, and in the air no odours that promised future good cheer. No merry young voices were chatting without and within in the exuberant anticipation of the Christmas joy.

Farfar Peter shrugged his shoulders, and had most uncomplimentary thoughts of the young husband, who had probably sold "piggie" whole, without leaving so much as a bit of good pork for that slip of a wife of his to busy herself with, like her neighbours.

Christmas Eve came like every other evening to Farfar Peter, in stillness and silence and dimness; for there was but one little "dip" burning in his small room. Farfar Peter sat in his great rush-bottomed chair, of home manufacture, and let his thoughts run back half drearily over the cheerful past, so strongly contrasted with the doleful present.

The door was suddenly burst open, and in another moment a strong boy had his arms round Farfar Peter's neck, and his head laid on the old man's shoulder as he cried out hilariously, "Dear Farfar, I had to come to see you! To think that I should get here just on Christmas Eve!" And then Hans sat down, and looked very hard at Farfar Peter.

There was no answering sparkle in the old man's

eye. His great chest heaved like a ground-swell after a storm at sea, and his mouth was tight closed. He had never been so moved before in all his life. Suddenly a new thought struck him as he looked at the boy's sailor clothes, and he said hurriedly, "You haven't run away, Hans?"

"No!" burst from Hans, proudly. "I told father and mother I wanted to go to sea, and they wouldn't hear of it at first; but when I said that you, Farfar, were at the bottom of it, and that I wanted to work my way across to you, for I couldn't think of your being here all winter alone, with never a boy to do a turn for you, they wheeled right round, and were so glad—you can't think how glad they were. Father kissed me. I don't think he ever kissed me before in his life. And mother, she cried, and pressed her hand on her heart, as if she was afraid it would jump out somehow, but she couldn't say anything just then. They fixed me up and got me off as soon as they could. I liked the sea right well, till we took a storm, and were near going to the bottom. A steamer picked us up, and carried us to England. They were ever so kind to me in the Swedish Sailors' Home in London. I was sick there quite a while, but now I am well, and safe and sound in old Sweden. Have you got anything to eat here, Farfar?" said the boy, most unsentimentally closing his story.

Farfar Peter pointed to the cupboard, and the amount of bread and cheese consumed that evening, and the amount of talk that accompanied the modest entertainment, could hardly be estimated.

They both slept charmingly until the early morning, when there was a sound that made Farfar Peter start up in his bed. "That's a baby, as sure as I'm alive!" he exclaimed. And a baby it was, to be sure.

There was a warm gush of pleasure in the old man's heart. It was so long since he had heard a child's voice. Hans was not much moved by the occurrence, but was glad to be waked, for he was impatient to look about everywhere and enjoy the old memories of the past. Some changes there were, which did not quite meet his approbation, for he felt as if he were in his own dear home, which belonged to him as much as ever. He finally put in his head at the cottage door. He did so want to see if all were as it used to be within. What a welcome sight that cheery face was! The young husband wanted a messenger to go after the doctor immediately, for it was plain that his wife was very ill.

All day there was something for Hans to do. He

was here and there and everywhere, and it was soon
known far and near in the neighbourhood that one of
" the boys " had come home to see Farfar Peter, and
so Hans was suddenly quite a hero in his way, and
there could hardly be a doubt that he would find
employment for the winter.

The baby thrived, but the mother continued ailing
and feeble. At last she was pronounced able to sit up
once more. There was no seat that was comfortable
for her in the cottage. Farfar's own big chair would
be just the thing, Hans was sure. He was now quite
at home with the neighbours on the other side of the
wall. The chair was lent at once, " with all the pleasure in the world."

A few days later, Hans came to the old man to ask
if he would mind having the door opened between his
room and the cottage, as the young mother thought it
would be homelike if he would look in upon her now
and then.

He did " look in " through the opened door, and took
his seat in his own big chair, which had been placed
beside the cheerful fire. The mother had the baby in
her arms, and Farfar Peter hung over it with a smile
on his great, rough-hewn face, and a blessing upon it
in his honest heart.

"I have thought a great deal about you, Farfar Peter, as I have lain here by myself," said the mother. "I don't feel we have done quite right by you, my husband and I. We almost forgot you were there alone, we were so happy together. Hans has told us all about you and that Irish family over in America, and how he had to come home to see how you were getting on. We haven't done right by you."

This was the beginning of a cheery time for Farfar Peter. He went in and out at the opened door as much as he pleased, and his great chair kept its place near the warm fire, always ready for him. Hans was such good help, the mother said—the handiest boy she ever saw; and she couldn't get on without him, and it really seemed best that they should all take their meals together.

How good the warm food tasted to Farfar Peter, and what a sunny atmosphere there seemed to be suddenly about him! Of course, he had to be at the baptism, and be godfather to the boy baby, and to drink—it must be told—some of the strong sweet stuff that was passed round, as if it were an essential part of the ceremony.

Old Peter and little Peter—for there were now two Peters in the cottage—were very good friends. It was even soon confidently affirmed that the wise baby looked contemplatively at the wooden leg, and evidently understood that Farfar Peter was not quite like other people.

"Not like other people!" That was what every soul in the red cottage thought about Farfar Peter. They honestly believed he was the nicest, the very best man in the world. He, good soul, could not forget the bitter thoughts he had cherished when he was all alone. He knew very well that he had his faults as well as other people.

IV.—CONCLUSION.

THE Irish family seemed celebrating a spring festival in their own peculiar manner. They were all out round their home as busy as bees in June. Old and young had each a basket or a box or a bag in which to gather most unseemly treasures. The refuse of the yard had evidently been classified, if not according to scientific principles. A special prey had been appointed for special individuals. Such a jolly time as they had with their scavenger work! Bridget beamed with satisfaction. Those Swedes should see that she could have her yard as clean as theirs if she only chose. Faith! she meant her windows, too, to shine so that it would hurt the neighbours' eyes to stare at them!

The sound of wheels brought the busy workers to a standstill. All eyes and mouths were simultaneously opened wide as a handsome carriage drove up to the door of the Swedes. The glasses were down, and a great, kindly old face looked eagerly out with the curious interest of a stranger. There was a panic in the Irish camp. Stina's counter-manœuvre was at once understood. "She was not going to let those Irish people beat her in showing respect for one's parents!" she had said, and they knew it as well as if they had heard her.

Hans got out first and held the carriage door open, and then Stina, and then the huge old man stood on her side-walk like a crippled but happy giant.

First Axel laid his hand on Farfar Peter's shoulder, then Farfar Peter bowed his grey head to the son's shoulder. They patted each other on the back, and then they kissed each other, and that part of the ceremony was over. Then Stina went through with the same process, while the children respectfully waited for their share in the family scene. Then there was the bobbing of short courtesies, a bowing and a kissing of the old man's hand, that the Irish family watched all agape and with evidently growing depression. The grand opening of the reception being over, all eyes on Pat's premises were turned towards the carriage to see the next guest.

The coachman leaned down towards Stina, put out his hand, then closed it again, and promptly drew up the reins and drove off, feeling a little queer at his heart from the scene he had witnessed, and most comfortable as to his pocket, where he had thrust the more than necessary payment from Stina, who had not forgotten the rules for *pour-boire* in her own country.

There was a long, triumphant howl from the Irish children, and then a joyous shout, "*They* haven't got any *granny!*"

The fat, little, white-capped grandmother was suddenly seized by the oldest boy, and then hand joined hand until there was a glad ring of the wild little youngsters who *had a granny*, and must exultingly celebrate their victory.

Round and round they capered and danced, the old woman being by no means the least active of the party, whether with her own will, or without it, no

observer could tell, but she evidently felt a joyous
pride in the unexpected ovation with which she was
so suddenly honoured.

Yes, Farfar Peter had really come to America to
cast in his lot for his last days with his kith and kin
in the far, strange land. He had fairly deserted
little Peter. The more he had loved that baby, the
more he longed to see all his own grandchildren over
the sea. Those Irish people had ventured on the
voyage, and why should not he, with a strong, know-
ing fellow like Hans to help him? He should still be
a Swede and have Sweden in his heart wherever he
went, and Sweden would be around him in the home;
for Swedes Axel and his family would always be, he
was sure.

Farfar Peter arrived with money in his pocket.
He had sold out. He was no longer "a reserve man,"
"an exception," but a grandfather who could pay his
way in hard cash and have his full right to a warm
seat in the chimney corner. A warm place in the
heart of his children and children's children he knew
he had already.

Farfar Peter had been sure of his welcome, but to
see all the dear faces once more, and hear the children
rejoicing about him, almost unmanned him, when he

remembered how he had sat and grumbled in the red cottage on the other side of the water.

Farfar Peter did have a Sweden about him, as he expected. There was always pea-soup with bacon on Thursday for dinner, and porridge of some kind for breakfast and supper on any or every day of the week, and the dear old hymns were sung morning and evening by the whole family, as if they were still in the land of Gustavus Adolphus. When Christmas-time came round again, there was an abundance of good cheer in the old fashion. The cod-fish was steeped in lye, then boiled till it looked like a white jelly, and piggie was king for more than a day, and the honoured centre of the culinary operations.

"Those foolish, thriftless Irish! A turkey, to be sure, for their worships, and not a bit of baking going on!" Stina had exclaimed as she saw a big turkey and some shining red apples being carried into the opposite house. Stina could not know that Pat was getting a Christmas present from the lady whose garden he "worked." She was wearing her long crape veil for her own honoured father, and could never see Pat with his little "pappy" beside him, helping a little when there was much to do in the garden, without a tear in her eye and a warm feeling in her heart towards the honest Irishman, who had thought it a charming thing to import his poor parents, that he might take care of them in their old age.

When Stina had arranged at nightfall the great dish of rice porridge for the Christmas Eve feast, and duly "criss-crossed" it with brown lines of odorous cinnamon, the children asked her why she had two such dishes this year. She gave them no answer, but a little later she sent one of the boys to give her greeting to the old couple over the way, and to say she had thought they might like some real Swedish Christmas fare. There were gingerbread babies, too, for the Irish children; and wonderful babies they were, all a kind of caricature of the fat little Irish grandmother.

Stina's messenger was met in the middle of the street by one of the boys from the opposite house, with a steaming pan of potatoes in one hand, and in the other a basket of the very red apples that had kept the turkey company.

The boys passed each other in silence, as the tacitly-accepted rules of war prescribed. The Swede was admitted. He bowed politely, delivered his message

and his good cheer, and disappeared before the astonished recipients could rally to return their thanks in
suitable form.

The bare-headed Irish boy came plunging into
Stina's room, but suddenly stopped short as he saw
the little lighted Christmas tree in the middle of the
white-covered festal table, and on each side of the
sparkling central ornament two high, branching,
home-made candlesticks, twined with gay-coloured,
fringed paper clipped by the children's busy hands.
The visitor opened his mouth wide, and stood in
speechless admiration. At last he managed to put
down his burden, bolting out, "Mammy sent 'em!
Grandpappy helped hoe the taters!" and then ab-
sconded with the deep conviction in his heart that
"there was no getting ahead of those Swedes."

When the sun was rising on Christmas morning,
the whole Swedish family set out joyously for the
early service in the Lutheran Chapel, where there
would be prayer and an address in their own dear
native tongue.

On the other side of the street, the Irish family
were coming out of the gate, bound on a similar
errand, but their goal was their own little church
with a golden cross on the spire.

How it happened, neither family ever knew, but they met in the middle of the street as by common consent. Then and there took place a cordial shaking of hands all round, and then a parting as prompt and silent as the greeting had been.

So peace and good-will between the opposite neighbours began on a Christmas morning.

ZACHARIAS AND THE BABY.

ZACHARIAS was schoolmaster, sacristan, and organist in the little parish of Moberg. He vaccinated the children, and was a common referee in all minor matters of law, as well as a kind of god-father general to "stand for a baby" at a pinch in any family whatsoever.

Zacharias lived in the school-house, where three rooms and a kitchen were, with firewood in abundance, a large part of the earthly reward of the teacher. The school-house stood by the church, as was proper where public education is super-intended by the ecclesiastical department and Luther's Catechism is a staple article in the instruction.

There was not a neighbour within call of the

school-house, but it was not a lonely place. On Sunday it was, of course, in the midst of the stir of the coming and going of the worshippers, and on week-days the air was ringing with the glad voices of the children.

Zacharias was known and addressed by the title of " Klockaren " (the bellman), though with the bells he personally had little to do, excepting that they were his silent companions when he betook himself to the belfry for a few moments of quiet.

Zacharias liked to call his wife Lotten, instead of Charlotte, and as " Klockaren's Lotten " she was distinctively known in the parish.

The Christmas vacation had begun. Lotten's five little girls had been uproarious all day ; but they were now in bed, and she was sitting by the kitchen fire-side, awaiting the return of her husband. She beguiled the time by wishing that everybody could have been made grown-up at once, instead of toddling about for years in other people's way. *She* was not fond of children, like Zacharias. She couldn't think of him as an old bachelor. And how could he have got on without her to see after him ?

Lotten sat quite still when she heard her husband stamping off the snow from his feet in the little storm-house. He soon came in with his overcoat on, and a stiff newspaper parcel in his arms.

Zacharias had lived so much among children that he was in many things like a child himself, and his wide-open blue eyes beamed with a kind of mild trustfulness and innocent surprise. Now he had a brisk manner that Lotten knew boded no good. She had often said that was "his way" when he expected her to find fault with him, and knew in his heart that he deserved it.

She did not leave her seat, but began, "Where have you been? Doing something for somebody, I suppose, as usual."

"Not exactly," he answered pleasantly; "but I have got a treasure for *us*. I've been at an auction."

"An auction!" exclaimed Lotten, dropping her knitting; "the worst place in the world for a man like you. You've bought something 'cheap,' I suppose, that we could well do without."

"I've been at another kind of auction," answered Zacharias; "I've been at the poorhouse. There were some of them there to be put out to the lowest bidder. Tailor Fred's father came first. Fred kept quiet. At last Tinker Tobias bid shamefully low, and then

Fred let the old man go creeping off with Tobias without a word, though he knew as well as I did what kind of a home the poor creature was going to. It made my heart ache to see it!"

"Now I am thankful!" said Lotten. "I was afraid we were to have the old man, and next to nothing to keep him on."

"I did think about it," said Zacharias; "I couldn't help it, but I did not see just how we could manage it. I did better. I brought you home a baby. See, here he is! A fine little fellow!" And the husband began to fumble at the bundle in his arms.

"A baby!" screamed Lotten; "and I tired of the sight of children! Our girls are a nice enough lot, to be sure, but then there are the school children, ramping and roaring inside and outside the house; and the babies to be vaccinated, too, screaming by the half-hour, with their sleeves rolled up to their shoulders; and all the little trash of the parish you've stood god-father to, coming to ask after 'guffar,' expecting at least to get a cake or a biscuit. And now you are bringing home a baby, to be sure!"

"Just hear how it was," said Zacharias patiently. "You know about poor little widow Maria, who strangled herself last week at the poorhouse—the doctor said the child hadn't any chance at all, unless it got into good hands."

"It isn't that baby?" screamed Lotten; "and its mother killed herself!"

"He said so to me, Lotten," continued Zacharias, as if he had not been interrupted. "Perhaps he minded what luck you had with Längy Han, bringing her up with the bottle, and she now the biggest and strongest of your girls. I didn't bid till old Mia at the mill came offering almost nothing. Then you may be sure I called out less, and *we* got him, Lotten. Just look at him. Such a fine little fellow!" And Zacharias tore off the paper.

"To bring home a baby in a paper, and risk him out in such a night, too! The law might take hold of you for it, Zacharias," said Lotten, critically regarding the new-comer, who was still sound asleep.

"As to the paper, it's the new notion that paper is warming," said Zacharias cheerfully, for he saw he was on the road to victory. "I put two good sheets of newspaper over my shoulders, under my over-coat, to-day to try it. There didn't seem to be enough on the baby, so I pulled down the papers, warm from being next to me, and had them snug round him in a moment. Never you fear about his taking cold.

When I was up among the Lapps, when I was a lad, I've seen them start out with a child not a week old to carry him a twelve-hour journey over the snow to be baptized. They'd be off with him one morning a heathen, and the next they'd be back with him a Christian, safe and sound. They are tough, babies are."

" Little you know about babies, Zacharias. I mind Hannah and those bottles ; I feel worn out just to think of 'em."

While Lotten was speaking, Zacharias was looking approvingly at the baby. Its fore-arms were placed across its tiny figure close to each other, and then its whole body was wound about with a long cotton bandage, until it was as stiff as a log of wood. The little stranger had a small, wrinkled face, as if he had seen much care and trouble. His close-fitting, white cap was without border, and of a coarse white shirting. He was not dressed becomingly for his first appearance in his new home. Now, waking from his quiet sleep, he set up a feeble yell, and wonderfully distorted his incipient features.

" He ought to have something to eat at once," said Lotten, the woman strongly asserting itself within her at the sound of the shrill little voice. " Not a bottle in the house that is not crammed full to the top with

berries.   What a pity you are such a temperance man,
Zacharias.   Now, in my father's house there were
always——"

"Here, you take him," interrupted Zacharias, dexterously freeing himself from the baby.   "I've got a
clean little bottle in the schoolroom that I was showing the boys experiments with the other day.   There's
a clean, soft bit of sponge, too.   We can stop it in
for a cork, and he'll take hold first-rate, I promise
you."

This arrangement was peaceably acceded to.   The
bottle was filled with milk and water, its outside
cleaned and polished, and then remorselessly thrust
into the midst of the porridge that had been kept hot
for Zacharias.   Now he knew the dance had begun.
Hereafter that bottle might at any time be found in
the most unexpected places—under his pillow to keep
warm, or in his drawer wrapped up in a pair of his
best woollen stockings, or thrust into the fresh-water
bucket to cool off.   He, too, knew something about
bringing up babies by hand.

When the little stranger was at last drawing
famously at the sponge, and choking occasionally in
the midst of his astonishing success, Zacharias said,
after a moment's serious reflection, "I had a little

brother once who died.   His name was Ernst, but we called him Esse.   We'll call the baby Esse."

"We won't call him after anybody that has died. That's bad luck," said Lotten, impetuously.

"A notion! only a foolish notion!" answered the husband with pleasant decision.   "Why, here I am a man, safe and sound, and I was named after the prophet Zacharias, who died more years ago than perhaps the pastor himself can reckon."

This was an argument Lotten was unable to answer, and she understood that Esse was to be the name of the baby.

II.

A FASHIONABLE home has been described as "a place in which to get ready to go somewhere else."   This description might well be applied to the parish of Moberg, for it had been regarded by many of its pastors as but the first round of the ladder that was to lead to clerical promotion and ultimate prosperity.

The parish was poor and the parsonage small.   The salary was inconsiderable, not more than two hundred dollars a year, a part of it paid in wood, hay, etc.—in short, as the Swedes say, *in natura*.   The present incumbent was a tall, stiff young man, with hair that swept in a black curve across his pale forehead like a raven's wing. His dark eyes had a gloomy, absent expression, as if he had not quite settled for himself all the difficult problems of human life. One thing was clear to him—his worldly position must be bettered; for he had a betrothed in another parsonage far away, and a home for her and with her was his bright aim for the future.

The pastor was just now laying a strong foundation stone for that home in a sermon he was writing, the best, he was sure, that he had ever penned. It could be preached, of course, to the poor, plain people about him, but should reach the ears of better judges on the first opportunity. In the midst of its closing periods he was half aware that there was an altercation going on in the room adjoining his little study. He heard, he was sure at last, the voice of his housekeeper, so called by common consent, though she was, in fact, maid-of-all-work for him, as she had been for several pastors before him—a trusty, devoted woman, who even toiled in his little garden with her own strong hands. She was, besides, guardian of the pastor's study door, that no one should unnecessarily break in upon him in the heat of composition. Now, however, she admitted a tall, strong child, with the introduction, "Here is Klockaren's Hannah, who must speak to the pastor."

The little girl so presented was left at once to state her own errand. Her plain, practical face wore a serious expression, as if she were sitting in church, and her opened hands hung stiffly down at her side, as if she had taken by command a gymnastic position. She dropped a courtesy, and as the pastor silently looked at her with an abstracted expression, clearly delivered her message :—

"Far * is out and mor thinks the baby is bad, and it would be best it was baptized at once, if the pastor could make it convenient."

The pastor looked dreamily at her, and answered, "Yes, child, I'll come as soon as I can."

Hannah "made her manners" again, and then took herself as quickly out of the house as if she had been followed by the unquiet spirit that dwelt in the large dark eyes that had looked into her own.

The study again silent, the pastor took up his pen, and wrote and corrected, changed and finally completed what he meant to be a most brilliant close to his sermon.

* *Far*, father; *mor*, mother.

Meanwhile, Lotten had laid the sick child on her own bed, over which she had cast a fair, knitted cover. A snow-white towel, of her own weaving, was brought out for the little table that was close at hand, and on it were placed a bowl of water and a fine napkin from the church stores. All things being so ready, she betook herself to the window to watch for the pastor.

Zacharias had meantime come in, and expressed his approval of Hannah's errand. He sat down on the bed, and looked long and quietly at the little creature who was lying there scarcely seeming to breathe.

Weary with restless waiting, Lotten turned at last to look at her husband. His face was very solemn, and tears stood in his eyes as he rose up. A bright drop shone on the cheek of the baby, and he tenderly wiped it away.

"Is he alive, Zacharias?" asked Lotten in a sudden fright.

"He's alive! Yes!" said Zacharias, and walking towards the window, he stood silently beside her.

A quarter of an hour later, the pastor was to be seen coming rapidly along the road that led to the school-house. Lotten was no longer watching for him. The five little girls, with their small shawls on, and

black kerchiefs over their heads, were sitting outside
the door, whither they had been banished by their
mother. The three older children were close to each
other on the upper step, with their clasped hands on
their laps, while the two little ones sat below them,
crying aloud, and lifting up their voices more de-
cidedly as they saw the pastor approaching. The
children rose and filed on each side of the steps,
courtesying simultaneously as he passed.

Zacharias and his wife received the pastor with the
usual signs of respect, but in silence. Lotten was the
first to speak. "He is dead, pastor! The baby is
dead!" she said impulsively, pointing to where the
child lay.

The bowl had been removed from the white-
covered table, and there now rested little Esse, the
pure napkin wrapped formally around him.

"He is really dead," said the pastor, and a heavy
shadow settled over his dark features.

"To think such a thing should have happened in
my house!" said Lotten, sobbing. "Such a dear
little baby as he was; so sweet and patient. We had
all got so fond of him in just these few days." And
she began to moan again.

"I can do nothing now. It is all over," said the

pastor. "I am very sorry," and he gloomily, almost sternly, took his leave. He was not the man to speak out the strong feeling that was surging at his heart.

There were many red-tape matters to be settled about little Esse. The doctor made out his certificate that the child had died a natural death. He had thought from the first "that baby would hardly pull through." The poorhouse authorities must see the dead body. Lotten, they said, must be paid a little for her trouble, and the parish must meet the small expenses of the funeral. "I can attend to all that," said Zacharias. "We loved little Esse, and we don't want any money for what concerns him."

These necessary details being settled, little Esse was removed on the table as he was to the school-room, so often before resounding with happy, childish voices.

Zacharias betook himself to the belfry. He wanted to be alone. The bell-tower stood outside the little church; it was a peculiar, independent structure, shaped much like a Chinese mandarin. There Zacharias was sure to be left to himself. The children were forbidden to intrude upon him, and Lotten had declared that she would never be seen there till the doves were banished. There was hardly foot-room for any one, indeed, for the doves had "found out a place for their nests" on the very floor of the belfry, and Zacharias now tiptoed around, not to step on a young brood or a pretty cluster of eggs, some mother-bird's treasure. There was no danger of the doves being ousted. Zacharias considered them a kind of church property, their young to be sacrificed occasionally for the good of the sick poor.

Zacharias did not put his foot on the treadle by which the bells were rung, nor did he lift the oil-cloth curtain that covered the Bible and psalm book that lay on a high shelf. He did not even look at what he called "Big Bingbong," the bell that was rung for ordinary church purposes, or "Little Bingbong," that was solemnly tolled on Saturday evening, or furiously rung for a fire alarm. Zacharias had his special names for the bells, as he had for many other things, but he was not now thinking of them or their names. He went straight to the open window, and looked out towards the golden sky of the winter sunset, as if he saw into the depths of a glory that could not soon pass away.

When Zacharias came down, he was evidently in a silent mood. He betook himself at once to the school-room, not apparently to see little Esse, though he did

look tenderly at the still, little form, and even measured
it carefully with the rule he always carried in his
pocket.   Carpentry and the catechism were favourite
departments of instruction with Zacharias.   A rough
bench with tools neatly arranged upon it stood
behind the schoolmaster's desk.   There he placed
a pine log he had brought in in his arms.

There was soon the sound of steady work in the
before silent room, while Zacharias hummed a psalm
tune, as he skilfully handled chisel and plane.   He
was up late that night, and busy every hour that
he could spare the next day.   There had soon
appeared a narrow, oval box, turning over at the top
in heavy scrolls at the ends, to be afterwards everywhere lightly carved with leafy devices.   It was a
casket for something precious he was making.   He did
not consider it exactly a coffin, though on the cover
stood "Little Esse" in pretty, raised letters, and on
each side of the name was the head of an angel between
two wings.   They were not likenesses of heavenly
visitants, but of his own little "Veevee," who
patiently posed before him as a model.   Veevee was
her father's favourite, and more like himself in
character than were any of the other sisters.   Zacharias called his five little girls, in every-day life,

after the pet names given by Swedish children to the fingers, though he had modified them a little to suit his purposes. Useful Längy Han represented the thimble-finger, and fair, golden-haired Veevee, as yet merely an ornament to the family, had the dainty little finger to thank for her sobriquet. Akin to the angels her father thought her, and her mother was privately much of the same opinion.

III.

THERE was a shadow over the red school-house by the church, during the days that little Esse was lying cold and white in the school-room. The children, however, had had their own quiet pleasure in dexterously cutting soft paper into openwork patterns to be laid round, in the hollow of the tiny casket, to circle about the silent baby with a wreath of winter frostwork. The morning for the funeral had come. The little girls were busily strewing a pathway of sprigs of bright spruce from the cottage door to Esse's last resting-place. It was a narrow path, for Zacharias himself was to carry the pretty casket he had so lovingly made. Later in the day, the

solemn train left the school-house. Zacharias went
first, decorously bearing his light burden. The clergy-
man followed. The children were pleased to see that
the pastor was wearing his gown and bands as if it
were a big funeral. The big bell, too, was tolling
out its solemn sound, and to its measured movements
the little girls in pairs went stiffly behind their
mother, who gently led the weeping Veevee. The
poorhouse officials brought up the rear, in a business
manner, quite unlike the spirit that pervaded the rest
of the procession.

The churchyard lay like a rolling white ocean,
with here and there a half-fallen wooden cross ap-
pearing above the snow, like the mark of some vessel
shipwrecked near an unkindly shore. One corner of
the enclosure was shut off from the rest by a light
paling. Here the usually-locked gate was now wide
open. Within, the snow lay level and fair, save
where one long hillock had been covered by evergreen
branches laid lightly over it. This was poor Maria's
grave, "without the pale." She had been the first to
find a resting-place in that dreaded spot set apart for
the last earthly home of the evil-doer.

The pastor was not surprised that the little pathway marked on the snow led through the churchyard

to the open gate. Zacharias had said to him at the house, "Little Esse had nobody that belonged to him among the people there in the graves, so I had a place made for him by his mother. It seemed less lonely-like, too, for both of them. I suppose it was no harm, pastor." The pastor had bowed his head, while a painful cloud swept over his face. He was in no mood for talking.

The dark group soon stood around the tiny open grave beside the green hillock. The pastor lightly cast the three shovelfuls of earth upon the lowered coffin, and then read in a hushed, hesitating voice the prescribed words: "Dust thou art, and to dust thou shalt return. Jesus Christ our Saviour shall raise thee up at the last day." Here he handed the book hastily to Zacharias, whispering, "You go on. I cannot."

Zacharias went solemnly through the Lord's prayer and the blessing, as the rubric prescribed for such cases. Then clasping his hands over the closed book, he said in clear, strong tones: "Suffer little children to come unto me, and forbid them not: for of such is the kingdom of heaven. Inasmuch as ye have done it unto one of the least of these my brethren, ye have done it unto me." Then he broke out into

singing with his beautiful voice a verse from a familiar hymn,——

"Where I lay me down to sleep,
Gentle Jesus watch will keep!
May I wake to Him adore,
Love and praise Him evermore!"

The five little girls joined in with their father at
the second line of the stanza, as they were accustomed
to do when he sang.   At the last two lines, a group
of the school children, who had gathered without the
churchyard's bounds, came in heartily, as if a choir
of angels were singing from afar of the joy in store
for little Esse.   There was a short, solemn, prayerful
silence, and the simple service was concluded.

Zacharias lingered a few moments to see that Esse's
grave was properly fashioned, to rise, covered with
green, as if cared for, as he lay beside his dead mother.

The pastor and the poorhouse officials must ac-
cept the invitation to enter the school-house cottage
with Lotten.   She apologized that she could only
give them a cup of coffee that was hot on the fire,
as Zacharias, they knew, was opposed to the good
old-fashioned way of taking something strong after
standing in the cold air in the churchyard.

The pastor soon took leave stiffly, and strode away
towards the parsonage.   Veevee, bareheaded, and her

light hair floating out behind her, sprang after him. She soon saw she could not overtake him, and called out eagerly, "Pastor! Pastor!" He turned and waited for the child. She dropped a humble courtesy, and said earnestly, "I must ask the pastor. Where is our little baby?—not what we put in the ground. Where is our little Esse?"

The pastor looked down at the sweet child-face uplifted questioningly towards him, and hastily answered, "The baby is in heaven with God and the holy angels!" His heart had spoken out of its abundance, let the head think what it would! The dark shadow swept away from his face, but the cloud stole over it again before he reached the parsonage.

Veevee came running into the cottage with the gladness of a messenger of mercy. She had dimly understood from her mother's moans and lamentations the real cause of her distress.

"The baby's in heaven!" the child said joyfully. "In heaven with God and the holy angels! The pastor said so, and he knows!"

The poorhouse officials shrugged their shoulders and went out suddenly with but a hasty, parting bow, while Lotten's face reflected something of the joy in little Veevee's.

That evening, when Zacharias and his wife were sitting together after the children had gone to bed, she said to him abruptly,—

"The more I think of it, the more I am sure the pastor did not mean what he said about Esse. It was just to comfort Veevee. That's plain to me! He would have told *me* so, if he could, I'm certain, for he must have known very well what lay on my heart, and it's my belief he felt about the same himself."

"Lotten," said Zacharias, slowly, "I haven't felt to speak of it before, but *I* baptized the baby."

"*You* baptized the baby ! When ? Where ?" said Lotten in astonishment.

"When I sat by him on the bed. I saw there was no time to be lost. The water was there, and I knew the good words, and I baptized the baby. His name is Esse."

"But I was in the room all the time. I didn't see you or hear you," said Lotten doubtfully.

"You were looking out of the window, Lotten, and I knew it. I didn't feel to talk about it, but I baptized the baby. I went over the holy words in my heart. I didn't speak 'em out. I couldn't see as it was needed. It would be known up there without my talking out, and that baby was baptized."

Lotten looked at her husband in speechless admiration. She remembered the bright drop on the face of the baby when Zacharias had been leaning over him. It was then! There had been a baptism in her house, and she had not known it! Her husband stood up before her in a new light. While she had fretted and watched and worried, "he—he had baptized the baby!"

One of what Lotten called her "confirmation chapters," which she had learned while she was "reading" with her old pastor (as one of the "children of the Lord's Supper") came to her mind: "Suffereth long and is kind, envieth not, vaunteth not itself, is not puffed up, doth not behave itself unseemly, is not easily provoked, thinketh no evil, seeketh not its own, rejoiceth not in iniquity, but rejoiceth in the truth." Here it seemed to her she had her Zacharias as plainly before her as if he had sat for his picture! And yet she had set herself up over him and not understood him all these years! She suddenly felt he was above her, and for the first time she reverenced him.

Not that Lotten said anything of all this to her husband when she broke silence. On the contrary, she gave him advice as usual. "You ought to tell this to the pastor, Zacharias," she said. "He will be as glad as I am, I am sure, that the baby was baptized."

" Esse can't stand in the church books and have a
certificate, for there wasn't any witness that knew
about it," said Zacharias.   " I thought of that, but I
was sure it wouldn't matter.   He don't need any
certificate *up there.*"

After a pause he added, " But if you really think I
ought to tell the pastor, I will."   He looked at his
big silver watch, and then said, " It's too late, now,
but I'll do it in the morning—if I'm alive," he
slowly added.

It was his usual way of speaking of what he meant
to do in the future; but now Lotten exclaimed,
" Don't talk so, Zacharias, as if you might drop off
any moment."   He seemed to her, in her new view
of his character, " too good to live."

Zacharias did live till morning, and started as soon
as it was proper for the parsonage.   His way led him
past the churchyard, and a childlike smile of pleasure
came over his face as he saw that, after the fresh snow-
fall of the night, the graves of Maria and her child lay
in shining whiteness.

Zacharias was received with unusual warmth at the
parsonage.   The housekeeper said she was right glad
to see him, for she was really worried about the
pastor.   His writing, that he'd been so taken up

with, he didn't touch any more. He had been walking, walking a good part of the night. There he was now—she had been watching him through the glass door—not eating a mouthful of his good breakfast, but staring down, as if he were looking into a coffin. Not but what he had been kind to her. He had never been so mild in his way to her before; that only made her the more worried about him. Zacharias had better go in at once, and not leave the pastor a chance to say he wouldn't see anybody.

Opening wide the door, the housekeeper said, " Here's ' Klockaren ' to speak to the pastor," and then disappeared as if frightened at her presumption.

Zacharias bowed ceremoniously. His bow was returned, but the pastor did not rise. Zacharias began at once. " I have come to speak about the baby's baptism."

The pastor's face grew darker, but Zacharias went on, " I didn't feel to speak of it, even to my wife, till last evening, and she said I'd better tell the pastor at once. *I* baptized the baby."

The pastor sprang up from his chair, and took his guest warmly by both hands.

" I thank you from my heart, Zacharias ! " he said. " You have taken a great load from my mind ! " He had never called him Zacharias before. It had always been formally " Klockaren."

" It was all right, in a way," said Zacharias. " The water was there ready, and I knew the good words, as often as I've heard them, and I baptized little Esse. I didn't feel to speak of it, and I didn't see as it was needed, as he was going where they'd know all about it. Lotten was in the room, but she was looking out of the window with her back to us, so there was no witness. I didn't feel to tell her what I was going to do. It's a serious thing, baptizing is, pastor."

" It is a more serious thing not to have baptized a baby when it seemed to depend on you. You have taken away from me a load I should have carried all the days of my life ! "

" It *was* a slip in the way of duty, pastor," said Zacharias, " and I don't wonder it lay heavy. It seemed to my mind what the church had ordered was important, and I did it. I couldn't quite think it would make any difference to little Esse where he was going. Those words, " Suffer little children," are pretty strong. It is not likely either you or I could keep little Esse out of heaven when the door has been set so wide open for the like of him."

" I thank you more than I can say, Zacharias,"

said the pastor, taking the big, honest hands again in his own. Then, changing the subject, he said warmly, "Come, sit down and take a cup of coffee with me. I think my breakfast would taste better if I had your good company.

Zacharias had had his breakfast, and had besides "something to do for somebody," and "could not stop."

So the pastor and Zacharias parted to meet often in the future, on a different footing from ever before. Zacharias, quite unknown to himself, became a kind of pastor for the parish as he had been for many of the flock before, and there was a strong bond between them.

When spring was busy with the birches, and hanging the "pussies" on the willows, the pastor obtained the position he had been "seeking," and took his young bride to just such a cozy home as had long flitted before his imagination.

"I do miss the pastor in many a way," said the housekeeper. "If it wasn't disrespectful, I might say I almost understood his sermons towards the last, and sometimes he seemed to be speaking just to me, which, of course, he wouldn't, standing up there, right in the pulpit."

When the early summer had strewed the meadows

with bright blossoms, Veevee and her sisters liked
to stand at the locked gate and try to cover little
Esse's grave with flowers. The eager hands were not
skilful at hitting a mark, and many a bunch of
buttercups or daisies fell scattering on the larger
mound farther on, and lay bright in the grass like
friendly stars. So tokens of love reached the mother
and child, as they lay there " without the pale."

A year had gone quickly round for busy Zacharias,
when he received a letter from the pastor, saying he
had now a son of his own, and had " written " Zacharias's name among the sponsors. " I know what you
think in that matter," wrote the pastor, " and I hope
you will remember my boy, too, sometimes when you
are up in the belfry alone. And, dear Zacharias,"
so the letter closed, " I must thank you again from
the bottom of my heart, and tell you I could not have
had the joy I have now with my own dear baby if
you had not baptized little Esse."

LIEUTENANT LONG.

OUR lieutenant was a little man, a handsome one, he thought, a perfect Adonis, if he had been but a few inches taller. He enjoyed his buttons. They were to him the stamp upon the coin. They put a special value on the silver so endorsed. If those shining rows could have been longer! If the breast that bore them could have been broader and higher up in the world!

Everybody liked the little lieutenant. He was a warm-hearted, sunny fellow, ready to lend a friendly hand where it was needed, and to laugh at other people's jokes as well as his own. He kept his private sorrow to himself, but sorrow it was, nevertheless, and weighed upon him sometimes like a nightmare. This secret trial grew almost insupportable when the lieutenant found his affections fixed on, in his estimation, the most charming women. His heart was often in his throat, wanting to speak out, when he was in her presence; but he did not even dare to ask her to dance, though she smiled benignantly upon him. What a figure they would cut together, to be sure! She was tall and slight, and as graceful as a young elm stirred by the summer breeze. Her beautifully-formed head sat on her slender neck like a queen on her throne; majestic, but too noble to be proud or disdainful. How he adored her! He would have died at her feet, if that could have done either of them any good. The lieutenant knew it would be a useless sacrifice, which would wound her tender heart.

The lieutenant kept up his outward cheerfulness as best he could, and continued to be a favourite everywhere, and a chosen guest in town and country. In cottage and castle he was welcome. He liked the castle, though, better than the seaside cottage or the suburban villa. In the noble old mansions he frequented he met the adored of his heart, who was ever to him queen on the premises, whatever other aspirants there might be to the throne.

The lieutenant had been invited to a hospitable

castle, dating back at least to the time of Gustavus Adolphus. The house, the host, the surroundings were perfect. There would be boating, and riding, and lawn-tennis, and nobody knows what other diversions, and, best of all, the "adorable she" would be there.

The lieutenant arrived as a guest towards the evening of a long summer day. He had a few hours of bliss, and then a cloud of gloom suddenly wrapped him round. He stood on a balcony, gazing out for a moment into the moonlight, to master his struggling, surging emotions. *She* was looking splendidly, and had smiled so sweetly on him—smiled down upon him, though—and he must quiet the pangs of his old trial, which now seemed so peculiarly hard to bear. As he stood leaning on the balustrade, but unseen from the window, he heard voices near him. He knew at once that one of the speakers was a certain giant of a young count who liked to linger near the fair one whom the lieutenant would fain appropriate and annex. She doubtless was the listener. Of whom could they be talking ? He could hear the masculine voice saying, " Yes, he is a nice little fellow. If he were only larger, there would be a man of him. He lacks vital power," continued the speaker, growing warm with his subject. " Vital power is the secret of growth. We see it in all nature. Diminish or destroy the vital power, and growth languishes, ceases, and then comes a slow living death, and, finally, the last struggle. If the little lieutenant had more vital power, life would have stretched out before him, a battle-field where he would doubtless have come off a glorious conqueror. But now—"

The lieutenant heard no more, for the speakers had retired from the window. He could see no one again that evening. He was humiliated, crushed!

At last he peeped in at the window. The drawing-room was deserted, the crowd of many guests had streamed out on to the lawn to enjoy the moonlight. He entered stealthily, surprised the host in an arm-chair yawning, murmured something about "fatigue after the journey, slight indisposition," and made his way to his room.

The house was full, and the little chamber in the tower had been assigned to the lieutenant. He looked about it in bitterness. It had been considered large enough for him !

" Kind nature's sweet restorer " soon put an end to his trials. A very short sleep it seemed to him, when he found himself staring at a small luminous circle on the wall opposite to him. He could not understand it. It looked like a little round door not larger than the palm of his hand. Temptingly ajar, it seemed to whisper of some strange secret within. He got up and began to examine into the mysterious phenomenon in the clear moonlight of the summer night. There was really a miniature door open, from which came a kind of phosphorescent light.

The lieutenant had heard that the tower room had long ago been the abode of an old alchemist, of whose "almost discoveries" wonderful stories had been told. He carefully examined the strange little cupboard that glowed with the mysterious light. Several small papers lay there, done up like apothecaries' powders. The first that he took up was labelled, "Vital power, for the promotion of growth. One powder will be all sufficient to make a small man as large as he can desire."

The lieutenant did not hesitate a moment. He seized the carafe by his bedside, poured out a glass of water, put the tasteless powder in his mouth, washed it down with a single draught, closed the little door, and lay down to sleep, to dream of the glad future and of the human angel whom he adored.

The morning sun seemed determined to open his

eyes while he was struggling against surprise. He roused himself and dressed with great rapidity. How astonished he was when he looked at himself in the Psyche-glass in the corner! What a transformation he had undergone! In vain he pulled his collar up, he pulled his coat-sleeves down, and twitched at the hem of his pantaloons. His neck, his wrists, and his ankles were ridiculously conspicuous. He had actually grown immensely in a single night. This would never do! To appear before his adorable looking like a schoolboy in his last year's clothes would surely alienate her for ever. He hastily packed his valise, rang, and announced to the servant that circumstances compelled him to leave the castle at an early hour. He left courteous parting words for the host, and, buoyant with hope, set off for his home, if his bachelor quarters could be so named.

With such a beginning, what might he not expect? He would be tall, very tall, before he ventured to speak out the deep wish of his heart. She should look up to him then, as she had smiled down to him before. And what a couple they would be: he, the model of manly strength; and she—she would be as she always had been, the most perfect of women!

Alas! the lieutenant soon found that he was growing like a tropical mushroom. The preparing of his

perpetually-ordered new suits of clothes kept his
tailor in a state of pleased astonishment. His sword
had been twice changed, and now a match to the
favourite weapon of the redoubtable Charles the
Twelfth was hanging at his side like a child's bauble.
He was growing as ridiculously tall as he had been
small before. His bright visions were fading away,
and he was wrapped in a new strange cloud of
gloom. What would be the end of this swift process?
It was evident that the dose of vital power had
ingrafted upon him the disease of a strange, unnatural
growth.

The sweet vision of a happy home, with his adorable
as the presiding deity, was a thing of the past. He
had another fate to meet, and he must meet it like
a man. He threw up his commission, and retired to
private life—a private life of the strictest seclusion.

Fortunately an independent income had early fallen
to the lieutenant's share as an inheritance. The very
cost of his long garments might have soon shipwrecked
a young officer dependent only on his pay. He had
to bow down now when he passed through any
ordinary door. It was plain he must build and have
a house in accordance with his proportions. He built.
A modest dwelling he called it, and the rooms were

few, but the ceilings were high enough for a banqueting hall, and the door was like the gate of a city. The floor of the second story had soon to be removed, and the lieutenant took his view of the outside world from what had been the "upper windows." He would not submit to looking all day at a common bed that suited his dimensions. His bed was like a row-boat, a form he fancied, for he had been a sporting man. Turned upside down by day, supported on stands, it looked like some strange old saurian creeping along the side of the one spacious apartment, reaching to the roof, which was now his habitat.

By day the lieutenant never went out. He prowled at night like a wild beast, having a special permit from high quarters to have the freedom of the city unquestioned by the police. He crept through archways, and stood in courts, and looked in at the attic windows. Unseen, he saw the poor in their homes. He drew from his deep pockets his full purse, and left money on window-seats, or hung on door handles, addressed to the supposed victims within. He peeped in to see the effects on the objects of his benevolence, and saw rioting and drunkenness when he had hoped to see thrift and innocent joy. He resolved to give magnificently, though anonymously, through the

Charity Organization Society that would help up the souls and bodies he had helped down. He would devote himself to the rescue of his fellows in danger, and so cheer his own loneliness and be a blessing to mankind.

The sound of a fire-bell brought Lieutenant Long, as he was now universally called, into the midst of the gathered crowd, indifferent to their scoffings or their amazement. From high windows he lifted down screaming women, as much afraid of him as of the fire. He had his arms sometimes as full of half-clad little children as if he were a big brother carrying off the contents of his sister's baby-house. He revelled in fire and smoke, and tasted some sweetness in the dangers he braved, and the warm gratitude of the astonished mortals he had saved from a dreadful death. He frequented the wharves by night. No careless, half-tipsy passenger, no desperate suicide, could find a watery grave while he was in their vicinity. He simply stepped into the water, took up the drenched victim, and carried him dripping to the police station, where the heroic lieutenant was well known as the universal "life-preserver."

Lieutenant Long had given orders at the post-office to have no letters delivered to him. He had been tormented by correspondents who urged him to put himself into the hands of an *impresario*, with whom he could travel "quite incognito," and amass a fortune that would make a Crœsus seem a poor man, and enable him to settle all socialistic difficulties in his dearly-loved native land. The floor of the lieutenant's apartment had often been strewn with a white shower, as of a sudden snow-fall, as he indignantly minced to bits the letters of such pertinacious meddlers. Now the postman never visited his door. He had no near relations to send him a kindly word, and *she*— *she* would never write. To have lost *her* was the bitterest drop in his bitter cup.

That inexhaustible vital power was ever pushing him onward and upward. He was forced to communicate with his fellows by the telephone that was always dangling at his side. He was still chronically hoarse with that long-past shouting to the little people far below him.

At last, in the midst of the wretched loneliness and depression of its high position, his head received a terrible shock. He bumped his forehead against the moon. She quivered, shook, and flew into atoms! the balance of nature was destroyed! Then came an immeasurable whiz, a rush, and a general crash, and

Lieutenant Long awoke Lieutenant Short in his little
bed in the tower room at the castle. It had all been
one horrible dream. He was a little man once more.
Joy filled him with deep emotion. Thankfulness—
real, serious thankfulness—was the overwhelming
feeling that pervaded his soul. It had been but a
torturing dream. He was not only awake from that
dream, but thoroughly awake to see all the blessings
which surrounded his earthly path—a competency,
an education, health, an honourable position in the
army of his dear native land, and *the* woman of her
sex, his peerless fair one, to silently adore, and,
perhaps, to speak out to some time, and let her make
her own choice in the matter.

The lieutenant was somewhat pale, somewhat more
serious than usual, when he appeared that day among
the congenial circle at the castle. The adored met
him with a pleasant smile that was to him like the
rosy hues of dawn, and seemed to whisper of a possible
coming day. What a walk he had with her that
morning! What a conversation on that seat under
the lindens!

She had said something about his early disappear-
ance the evening before and his pallor when he entered
the breakfast-room, with such a sweet, womanly

interest that he could not help telling her all about
his terrible dream and his joy on awaking.

She looked kindly at him, and said, "I could never
have supposed that you would really care about not
being taller when you are such a favourite with
everybody. And yet I ought to understand something
about such feelings. I have been so very sensitive
about being so tall. When I was a young girl it was
really a mortification to me, and I made myself quite
unhappy about it, and could not bear to go anywhere
among strangers. Somehow, it came to me later that,
after all, I had just the kind of soul-case that had
been given to me for a little while, to be soon laid
aside, and that it was of trifling consequence whether
I was tall or short, if I could only live right and try
to make other people happy."

A sweet blush stole up her fair cheek, and called
down the long lashes that sank to meet it.

"You *could* make *me very* happy," said the lieutenant impulsively. "Do you mind my being so very
small? *Could* you ever care for a little man like me,
who loves you with his whole heart?"

The lieutenant and his adorable did not walk back
to the castle arm in arm, but they appeared side by
side, and looking so supremely happy that their glad

faces told the whole story, which had indeed been more than half suspected before. The world might laugh if they pleased, but the lieutenant knew that each was the blessed complement of the other, and that they were to taste the best of earthly joys—a happy married life.

THE "POVER ONTOO."

"THREE small rooms and a kitchen!" These
were all the domain of a middle-aged English
lady, who had been served in what she called "dear
Hindia" by a score of dark-skinned natives.

In India her husband had made his fortune. To
Sweden he came to increase it; but the arithmetical
process proved a sad kind of reduction, descending
instead of ascending, and ended in a discreditable
failure. The crash and the crush were domestic as
well as financial, and Mrs. Brown soon found herself
a penniless widow. Left of the past, she had a brooch
of tiger claws heavily set in gold, a small service of
India china, and her daughter Aurelia. Aurelia
Brown's occupation as a migrating teacher in one of
the smallest of Swedish small cities was to her mother
a painful daily reminder of her own broken fortunes.
That her child should be giving lessons in English to

stupid children, who ought to know by nature the best language in the world, was too humiliating.

Aurelia Brown was a tall, pale girl, stiff in person, and so accustomed to self-restraint and self-denial that it had become impossible for her to be natural under any circumstances whatever. Her face, with its regular features, was almost expressionless, and her light-blue eyes told only of patience without hope.

Free of words, florid in colouring, and hasty in temper, Mrs. Brown had her daughter in as perfect subjection as if she had been a dark-skinned Hindu.

There was no servant in the Brown establishment. In the kitchen Aurelia had her hours of retirement; but Mrs. Brown never asked what she did then, and pretended not to know. Once comfortably established on the sofa, with her gouty foot made comfortable, she never moved until she was told that dinner was served in the tiny dining-room. The food prepared for her was always dainty and appetizing. Aurelia's lessons seemed to prevent her taking her meals with her mother, for when she announced that a meal was ready, she always had her bonnet on, and seemed in a hurry. Of her private sacrifices and abstemious diet no one had an inkling but the old woman who brought up the wood and made the fires in the primitive establishment. That Aurelia should take her meals in the scene of her culinary functions had become accepted as a matter of course.

One of Aurelia's pupils was an enthusiastic damsel, a certain Fröken Flit,* who had become much interested in benevolent enterprises, and had a special philanthropic scheme of her own. She embroidered, painted, and carved diligently, and, finally, had a private bazaar that produced brilliant results. The fund being so established, Fröken Flit added liberally to it from her own private purse, and begged faithfully from richer friends whom she had interested in her undertaking. At last she had fairly in order a long, low Swedish house. "Pauvres Honteux" stood in large golden letters on the front of the building, that no one need be ignorant of the purpose to which it was devoted. It was an asylum for ladies who had seen better days, a kind of poverty that appealed particularly to the sympathies of Fröken Flit. She had secured three inmates. There was accommodation for a fourth, and for this member of her quartette she was on the sharp lookout.

At this juncture the small city was one morning stirred by the report of a sad occurrence that occa-

* Miss Flit.

sioned many exclamations of astonishment, regret, and pity. The pity was for " poor Mrs. Brown."

Wholesale-dealer Johansson was dead! He had gone off suddenly in an apoplectic fit. The whole town soon knew that he had the day before sold the house where Mrs. Brown had had her apartment for many years free of rent, on the strength of Johansson's friendship for her husband. She had really come to think the whole thing quite natural, and calling as little for gratitude as any other perennial blessing, from any source whatsoever.

Fröken Flit's eyes glowed. Her sensations were by no means altogether painful. Perhaps Mrs. Brown would be the fourth inmate of the " Pauvres Honteux !" She would see her at once, state the thing delicately, and set the poor lady's mind promptly at ease as to her future.

Fröken Flit supposed, of course, that Mrs. Brown could speak Swedish after so long a residence in her adopted country, and possibly she might, if she had chosen to try. As it was, she had shrunk from meeting the outer world since her misfortunes, and had had a good excuse for her retirement in her lame foot, and so lived quietly in a kind of England of her own making. When she had intercourse with Swedes she carried it on through Aurelia, as an interpreter, such conversations always being, of course, for the mother, philological instructions upon the most modern methods.   Mrs. Brown had so far profited by them that she had been known to suggest to Aurelia the most apt translation of a Swedish word, when the weary girl was for the moment puzzled.   Yet to really speak Swedish Mrs. Brown never tried.   She spoke English, and considered it the fault of her interlocutor if she were not understood.

Fröken Flit found Mrs. Brown alone, and redder than usual, with traces of excitement and emotion in her face.

"Good morning!" said Mrs. Brown to the round-faced, cheery-looking young miss who came in upon her in the midst of her painful meditations.

"Good morning!" was the reply, in a pleasant voice, with a strong Swedish accent.

Aurelia's lessons were carried on in the good old-fashioned way.   She would have as soon thought of giving out the dislocated parts of a verb piecemeal to her pupils, like the bits of a dissected map, as of setting before her mother broken bits of china, instead of the dear India service.

Mrs. Brown had taken in at a glance the whole

faultless attire of the guest, with a rapid valuation of its cost, and courteously waved her to a seat, while she said, " 'Ave the goodness to be seated. *Hi'm hoccupied* this morning. Should be glad to 'ear your *herrand at once. Hi* don't speak Swedish."

Fröken Flit was one of Aurelia's most diligent pupils, and specially skilful in grammatical gymnastics, but now only the list of the parts of the English irregular verbs came into her mind, and she broke out in Swedish, " I can't make myself understood in English. I have thought—I have hoped—it has struck me — unfortunate circumstances — sudden death." It was not easy to say even in her own language what she had intended. Finally in despair she pointed towards the low building, which Mrs. Brown could see where she sat. It was the " Pauvres Honteux," and Mrs. Brown at once understood the visit and its meaning.

" *Hi* 'aven't come to that ! " she said, with apparently a dangerous attack of rush of blood to the head. " The Pover Ontoo, *h*indeed ! " and her handkerchief went to her eyes.

The visitor took the hand of the English lady, gave her some gentle, affectionate pats, administered a bath of eau-de-Cologne from the bottle on the table, and

finally left her hostess apparently more calm, but
plainly not in the mood for further conversation.

Alas for Aurelia that the "Pover Ontoo" had ever
been mentioned! It was the burden of her mother's
continual wail during the days that followed the
death of Wholesale-dealer Johansson. The "Pover
Ontoo" Aurelia really believed was the only safe
harbour her mother could hope for in the stormy
times that had come to her. She herself could see no
prospect of doing more in the future than she had
done in the past. There her mother could be at least
comfortable, and well cared for. "And Fröken Flit
was such a cheery person, her visits would give
mamma real pleasure. She could entertain anybody,"
an art in which Aurelia felt that she was peculiarly
deficient.

The "Pover Ontoo" proved but a *château en
Espagne* as far as Mrs. Brown's personal occupancy
was concerned.

Wholesale-dealer Johansson's funeral obsequies
having been pompously celebrated, Mrs. Brown received a legal paper in which it was fully stated that
the house where she lived had truly been sold the
day before the death of her husband's friend, but a
condition of the sale had been that Mrs. Brown should

"retain rent free during her natural life the apartment she now occupied, with every right of a paying
tenant."

In her joy at the announcement Mrs. Brown
actually kissed Aurelia, a process very unusual in that
establishment. Aurelia actually blushed, and tears
filled her eyes, it was so unexpected. She was suddenly prompted to say, " I should have missed you
sadly, dear mamma! I am so glad you are not going
to the Pauvres Honteux, and that we can still live
together. I do not see how I could have gotten on
without you ! "

What would have become of Aurelia if her mother
had gone to the " Pover Ontoo " had never troubled
the mind of Mrs. Brown. A dim vision of her own
selfishness forced its way to her heart, and found vent
in the words, " Poor Aurelia ! But *H*i'm not going
to the Pover Ontoo ! That's *h*a comfort ! They must
'ave *h*a dull time there. Suppose we *h*ask the poor
things to take *h*an *h*afternoon tea with *h*us. They'd
like *h*it, *H*i'm sure."

The ladies at the " Pover Ontoo " were duly invited to afternoon tea with Mrs. Brown. They came
at the time appointed, all three of them, and made
their *entrée* according to the established rules of pre-

cedence among them. The tall fat lady, and the tall
thin lady, whose pretensions had been ascertained to
be quite on a par, were accounted peers, and they
came over the threshold side by side, and moved
regularly forward, keeping step as if they were
on military drill. They were welcomed in English,
Aurelia translating to them as if she were an auto-
maton and had no part to play as a daughter of the
house. Mrs. Brown waved her hand towards the
sofa. The guests stood still a second in front of the
seat of honour, then the fat lady settled herself into
the right-hand corner, and the tall thin lady, with
a slight sniff, contented herself with the more ignoble
left. Meanwhile a little, delicately-formed lady, who
had entered quite naturally behind them, stood quietly
to be welcomed when her proper turn came. She
was waved to a small chair adapted to her physical
proportions, and took it with a kindly expression
almost amounting to a smile. Then began a fire of
conversation between the hostess and the sofa ladies.
When one finished, the other took up the word, and
Aurelia was kept busy translating for the loquacious
party. At last some question was added that neither
of the military guests could answer, and one of them
condescended to say, " Perhaps Maria knows."

Of course Mrs. Brown addressed the same question to the other stranger in English, Aurelia translating.

" Maria " answered in the sweetest voice imaginable, and in faultless English.

Here was a thorough defeat for the military party. Mrs. Brown forgot them entirely, and they would make no reply to Aurelia's poor efforts to entertain them.

" Maria " and Mrs. Brown were soon in an animated conversation. " Maria " was English, but had lived many years in Sweden, as she had married a Swede.

All the hard look went out of the face of Mrs. Brown. One could almost see something of her early traditional beauty. Her mouth seemed to grow smaller, and an expression of restful satisfaction had taken the place of the habitual discontent. Aurelia looked on in pleased astonishment. She seemed to realize for the first time what it had been to her mother to be cut off for so many years from free talk with a guest in her own tongue. Kept a prisoner by her lame foot, away from the outer world, with but her silent daughter for her companion, Mrs. Brown had been leading an unnatural life. Not that it all presented itself so clearly to Aurelia, but she felt warm and loving towards her mother, instead of dutiful and

patient and self-sacrificing. There was a sunny
atmosphere in the little room that the sofa ladies
could not cloud, and the time seemed to Mrs. Brown
to have been all too short when the guests rose to
take leave.

"Maria" had had a fund of pleasant talk on which
to draw for the amusement of her hostess, and cheerfully promised, at the urgent request of Mrs. Brown,
to look in often on her countrywoman to chat about
"dear England." She did not say that threatening
blindness made her unable to use her eyes, and she
had therefore always time at her command.

The exit was made in the same order, and with
even more dignity than the *entrée*.

"Maria" went out last, as if she did not notice
there was anybody before her, turning to give Mrs.
Brown a friendly smile as she disappeared.

"That was a visitor *h*after my mind," said Mrs.
Brown to Aurelia. "*H*it would not 'ave been so bad
at the Pover Ontoo, *h*after *h*all!"

One day Mrs. Brown saw a carriage with outriders
standing before the "Pover Ontoo." Of course she
watched to see what was going to happen. A gentle-
man got out and entered the low house with a quick,
eager step. "Maria" met him in the vestibule with

a fond embrace. When "Maria" made her next call on Mrs. Brown (her calls were frequent now) she did not allude to the circumstance, but Mrs. Brown must have the whole occurrence explained. The visitor was the new English ambassador in Stockholm, who had heard all about "Maria," a beloved cousin of his youth, and had come to ask her to be a member of his family circle, and of the society which he was sure she would adorn ; but she had refused.

"*H*and why didn't you go, Maria?" asked Mrs. Brown in surprise.

"Why, Eliza," was the reply, "I am so contented at the Pauvres Honteux, I should dread a change. My cousin will do something towards permanently endowing the institution, and then I want to stay. I have seen too much sorrow to want to begin life again in a gay household. I should be sorry, too, to leave you, Eliza. We can be much to each other in this strange land."

"*H*it don't seem like a strange land, now *H*i know you," said Mrs. Brown. "*H*and *H*aurelia is so *h*altered. She's quite waked *h*up. *H*i can't say but what she did *h*everything for me before, but now *H*i really believe she loves me. *H*i 'ave found *h*out *H*i 'ave the very best daughter *h*in the world. To

think *h*all this should have come *h*about since that
dreadful time when *H*i thought *H*i should 'ave to go
to the Pover Ontoo where *you* live, Maria, *h*and that
*H*i should get to calling people as old as we are by
their first names, as the Swedes do! Well, they do
'ave *their* good ways, *h*if they can't speak *H*inglish."

"SUPPOSE YOU TAKE HER!"

———

THE pastor was out on a parochial visit. He was a portly man, who looked as if he had tasted prosperity, and had found it pleasant. Swedish birches were over his head, and the green meadows of Sweden were about him.

The two-wheeled gig rolled rapidly along the narrow road, the reins were in skilful hands, and the pastor could give himself up to his own meditations. They seemed to be of a cheerful nature, though his errand was not an agreeable one.

At the red cottage, whither the pastor was bound, he had been rudely repulsed on his last visit. The surly shoemaker who called the dreary place his home had curtly said, "The only truth I hold to is, Poor folks have a hard time, and rich folks take it easy. I don't need a priest to make me see that." Then the

door was opened wide to indicate the best course for
an unwelcome guest.

The shoemaker, having so defined his position, was
to be seen thereafter lounging about on Sunday morning in his leathern apron, or sitting on his doorstep
working at his trade, while the church bells were
ringing.

The rebellious parishioner had been "severely let
alone;" but now it was reported that his purse was
empty, his wife was lying seriously ill, and there was
real distress in the crowded home. His friends of
"The Society," whose programme was simply, "The
ups should be down, and the downs up," had shown no
inclination to help him in his difficulties. The pastor
had suddenly decided to look into the matter himself.

The opening of the cottage door into the one room
of which its interior consisted made all clear at a
glance. The shoemaker sat by a little window, with
a row of disabled shoes on the sill. Around him the
floor was strewn with bits of leather, and beside him,
on his bench, were the parts of the pair of coarse boots
which he was busily making.

A mass of tumbled dark hair lay on a ragged pillow, and a pale, sharp face peered at the pastor from
the poor bed. The thin lower lip took a sudden out-

ward pout, and then there was a contemptuous drawing up of the pinched nostrils.

The blunt district doctor stood by the sick woman, trying in vain to make himself heard. A bow-legged baby was toddling about screaming, and four noisy boys were storming around, collecting their books before starting for school. The boys promptly disappeared at the sight of the pastor, and then the doctor, looking disapprovingly about him, said to the shoemaker,—

"There's no hope for her in such a place as this. Good care, good food, and good wine might possibly bring her round." Then with a second bow to the pastor he went out.

"She might get in at the poorhouse," said the pastor, doubtfully.

"And die there!" broke in the shoemaker. "She might as well die where she is. You seem to have good room up at the parsonage, suppose *you* take her!" he added, with an ugly leer. Then he rudely held the door open, and roughly scraped his foot, as a clear intimation that he considered the interview over. The ungentle hint was promptly acted upon.

*   *   *   *   *

At the parsonage there was a stir of eager prepar-

ation.   Two young guests had been ousted from the
pleasant spare-room, and consigned to more narrow
quarters.   There was a bustle of sweeping and airing
and adorning, as if the queen were expected.   The
clean sheets had home-made lace let in below the hem,
and the pillow-cases were in the same style.   One bed
had been "telescoped together," and now stood like a
square frosted cake, shut in by a brown railing, while
the other at full length was turned open, just ready
for an occupant.   The paper window shades were
half lowered, and the netted curtains tied back with
pretty ribbons.   Bright strips of home-made carpeting
marked gay paths across the white floor, and on a
little table beside the bed lay Bible and psalm book,
with cheerful marks peeping out from the gilt
edges.

These preparations were hardly completed when the
guest arrived.   A light spring wagon had been filled
with hay, and there, on a feather bed and wrapped in
soft blankets, the shoemaker's wife had had an easy
ride to the parsonage.   She was gently carried to the
pleasant room, and dressed in the clean night-dress
and cap that lay ready for her.

She sank back on the pillows astounded and ex-
hausted.   She closed her eyes, and scarcely heard when

a step drew near the bed. A careful hand raised her and bolstered her up, while a tray, most tempting in appearance, was set beside her. She did not fairly look up even when a delicate bit of omelet was put into her mouth. Then came a taste of dainty biscuit, and then a great spoonful of wine that seemed to warm her through in a moment. She peeped under her drooped lashes, and saw a sweet, kind young face bending over her. The patient shut her eyes tight, and let the pleasant play go on, till she was really revived by the strengthening meal, and was quietly lowered from her half-sitting posture.

What a nice sleep followed! When the sick woman awoke, she was glad to find herself alone. She looked about her in wondering satisfaction, and could hardly believe she was not in a pleasing dream.

The door opened, and in came a little rosy-cheeked girl, who "tipped" to the bedside, dropped a courtesy, and put down on the table a beautiful yellow rose, in a vase of shining glass. There was no acknowledgment of the pretty gift, and the little visitor disappeared, to leave behind her, it seemed, a fragrant silence.

The eyes of the shoemaker's wife fell on the ornamented sheet, so pure and smooth. On it lay one of her thin, brown, not over-clean hands. She hid it hastily under the covering and turned over, to fall into a long and peaceful sleep.

Of course the doctor had said the patient must keep on with the quinine and turpentine pills, for she had pneumonia; but it was on her strong constitution and good care that he really founded his hopes for her recovery.

The white-covered trays that came to her bedside so often with tempting food on pretty china were always to her an agreeable surprise, and she submitted to being fed like a baby, though she knew very well she could have used her own hands if she had dared to show them against the background of sweet cleanness about her.

Morning came after a better night than the patient had yet had. Even her cough had grown less troublesome in the pure air she was breathing. She had enjoyed such a nice breakfast as she could never have imagined, when a tall, slight little girl came in quietly and shyly with a large comb in her hand. " May I fix your hair ? " she asked gently, after a polite courtesy. " I can get up behind you in the bed as I do for mamma when she is sick."

The patient meekly submitted. She had wondered

how her tangled hair was looking in her present
surroundings. How skilfully her poor, thin locks
were divided down the middle behind, and combed
out by the small, practised hands! It was really a
soothing pleasure to have those light fingers working
about the weary head. And how sweet the air grew
from the refreshing mixture the little hairdresser had
in her pretty bottle! This part of the toilet completed,
the volunteer maid proceeded to prepare some soft
water, scented with eau de Cologne, and to bathe
unasked the face of the patient, and to smooth the
black bands of hair above the pale forehead.

The shoemaker's wife looked into the shy, sweet
face of her little friend, and then ventured to say,
" It would be good to get my hands into the water too."

The bright-flowered bowl was replenished, and most
skilfully the brown hands were singly placed in the
warm water and tenderly sponged by the small, soft
fingers.

" Do keep on a little longer," whispered the patient,
soothed by the gentle bathing. She had other reasons
for not wishing the process to be too short. It was
over at last, and the poor hands had been wiped with
a fluffy towel as carefully as if they had been vases of
delicate porcelain.

The sick woman lay and looked with mild satisfaction at her wasted, big-knuckled hands as they rested on the white sheet, until she fell into a calm, untroubled sleep.

The patient woke suddenly. She almost thought she must be in heaven ! There was a sweet sound of organ music, such as she had heard at church in her girlhood. (Had the door been left purposely ajar ?) She could follow the words of a well-known hymn,—

> " The name of Jesus, comfort sure ;
> In all our need a port secure !
> Without the storm, without the waves,
> When Jesus speaks, when Jesus saves."

The poor sick woman had expected to be " preached to and prayed with " at the parsonage, and had resolved to show how little such things moved her. She had even been determined " to speak her mind " about such old-fashioned notions, if she had strength enough to do it.

Tears now stood in her eyes. She had to wipe them away more than once before the singing was over.

After a day or two, the shoemaker himself made his appearance at the parsonage, saying gruffly he supposed he might see his own wife if she were not too stuck-up to speak to him.

One of the little girls who had come out to meet him showed him politely to the sick-room, opened the door, and left him. Such a paradise as he seemed to be looking into! He took off his rough shoes, though he knew he had holes in his stockings. He had seen that there was no one but his wife to notice them. There she lay in the clean bed, with her hair smoothed, and her neatly-bordered cap framing her pale face.

He stopped and gave her a long look of pleased wonder. His mouth twitched till his bushy beard and mustaches moved like a thicket full of frightened hares.

"Why, Margit," he exclaimed, "what have they done to you? You have almost got back your good looks."

"They've been so kind to me," she said, in a low voice.

"And you've stood up for the right, and let them know there was no use trying to convert you, and bring you round to their way of thinking?" said the shoemaker, trying to feel belligerent.

"The pastor and the missis have looked in friendly at me, but it's been mostly the young folks that have been about me," she said. "They have nursed me

day and night, as if I was just as sweet as one of them.
I couldn't contrary them when they never said a word
of preaching to me.   I couldn't begin, you see."

The shoemaker looked round the pleasant room and
took it all in.   "And you live here, and they take care
of you ?   Now I am beat !   You'll do !   I may as well
go."   And he went out, shuffled on his clumsy shoes,
and left the house without speaking to anybody.

After some weeks, devoted care had done its work,
and the shoemaker's wife was going home.   The
thought was not altogether agreeable to her, but the
time had come and it must be done.   At the door of
the red cottage she was helped out.   The pastor's son
had driven her in the little gig.   He left her on the
doorstep, and the shoemaker took her by the hand.
The carriage did linger a moment, for the pastor's son
wanted to see what went on within.

There was a start of joyous surprise.   No leather
clippings littered the well-scoured floor, now sprinkled
with tiny sprigs of spicy juniper.   A little room seemed
to have been blown out, like a bubble, from where the
shoemaker had generally sat at his work, and there
were all his belongings, while the rest of the room
was the picture of neatness.   A big rocking-chair
stood by the clean bed, and near it a little table was

set out, with coffee and a strong luncheon for two.
On a big cake of saffron bread, white sugared letters
said plainly, "Welcome Home!" The wheels rolled
away, and the shoemaker and his wife were left to
themselves, until the children should be brought home
by the neighbours, who had kindly cared for them
during the absence of their mother.

The young guests at the parsonage had a vivid
description of the happy return to the cottage. They
had come to the pastor's to prepare for confirmation.
They had now the "Larger Catechism" stamped into
their memories, with much information as to the
fathers of the church and a clear outline of Bible
history. They had learned, too, that a Christian must
not only be sound in doctrine and strong in faith, but
full of the love that delights to minister to the suffer-
ing children of men. It had been a joy to them, out
of their abundance, to make the shoemaker's cottage
more like a home, and wise heads and willing hands
had helped them to do it.

When Sunday came, the shoemaker and his wife
and all their children (even the bow-legged baby) were
at church. The father of the family had said, "It is
my way of thinking that if the pastor, who has been
so kind to us, chooses to stand up there and preach,

we might at least be willing to sit comfortable and hear him, if we do know it won't do us any good." The wife said nothing, but the great Physician knew that she had come to him to be healed.

[The sketch given above is drawn from real life in a Swedish parsonage.]

THE END.